W9-AZT-476

Where Do
All the
Birds Go?

Where Do All the Birds Go?

Tracey Lewis

E. P. Dutton   New York

First published in the United States in 1988
by E. P. Dutton,
2 Park Avenue, New York, N.Y. 10016,
a division of NAL Penguin Inc.

Originally published in Great Britain in 1987
by Macdonald & Company (Publishers) Ltd

Printed in Hong Kong by South China Printing Co.
First American Edition   OBE
10 9 8 7 6 5 4 3 2 1

Library of Congress Cataloging-in-Publication Data
Lewis, Tracey.
   Where do all the birds go?/Tracey Lewis.
      p.   cm.
   Summary: Answers questions, in simple text and illustrations,
   about where animals go in the autumn.
      ISBN 0-525-44427-0
      1. Animals—Wintering—Juvenile literature.   2. Autumn—Juvenile
   literature.   [1. Autumn.   2. Animals—Habits and behavior.]
   I. Title.
   QL753.L48   1988   88-3743
   591.54'3—dc19   CIP
                                                               AC

to Rob

In the autumn, where do
all the birds go?

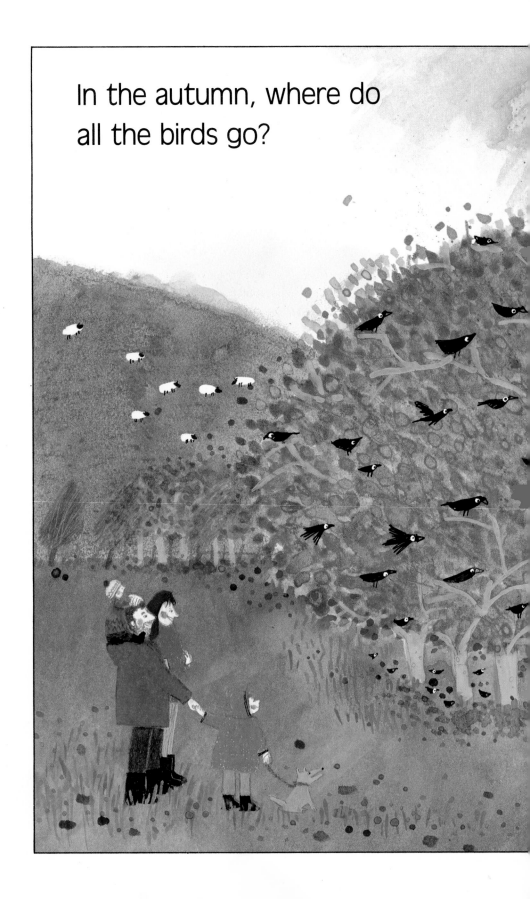

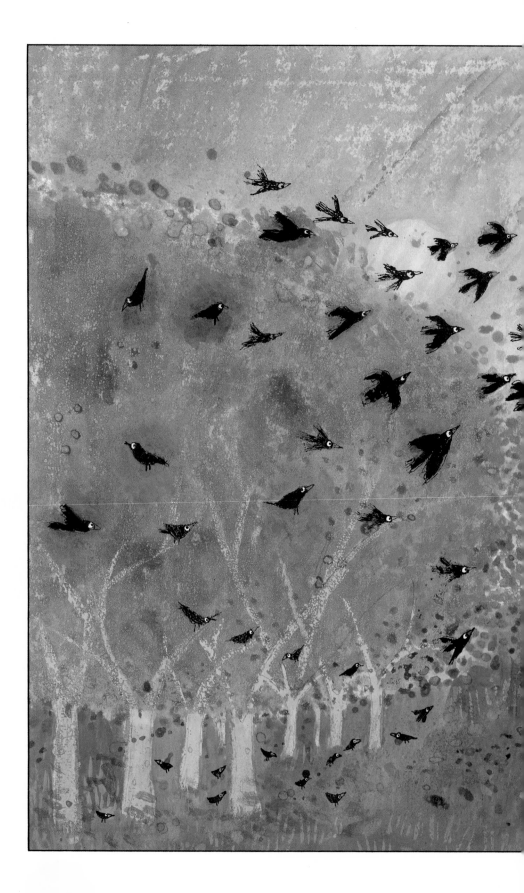

They fly away to warmer places.

Where do all the tortoises go?

They snuggle down
in cardboard boxes
filled with straw.

Where do all the squirrels go?

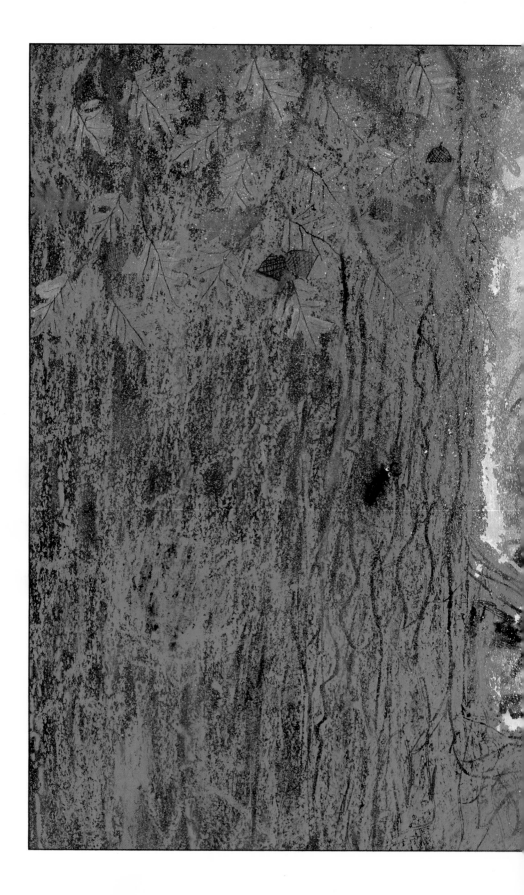

They settle down to sleep
in their treetop nests.

Where do all the horses go?

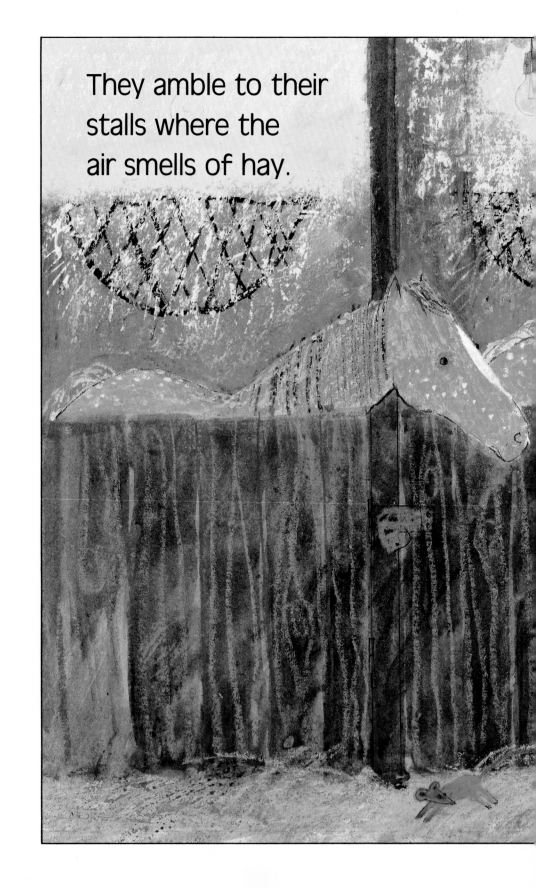

They amble to their
stalls where the
air smells of hay.

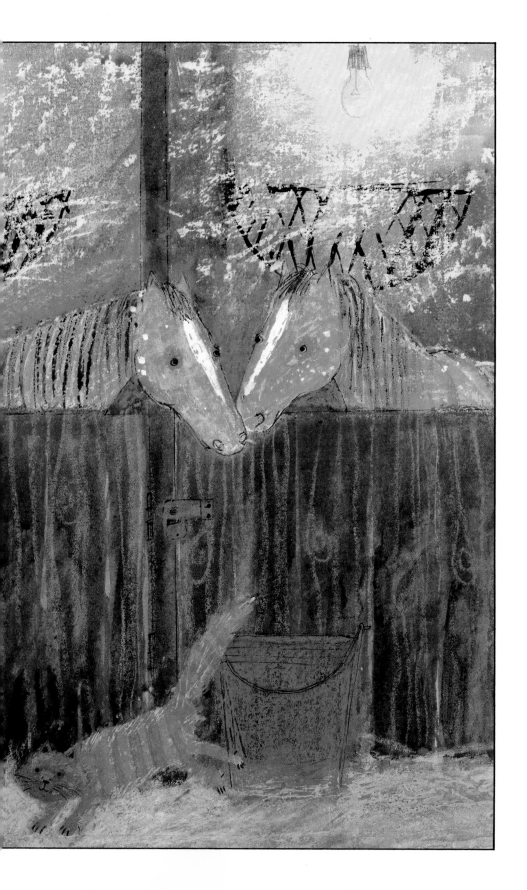

Where do all the house mice go?

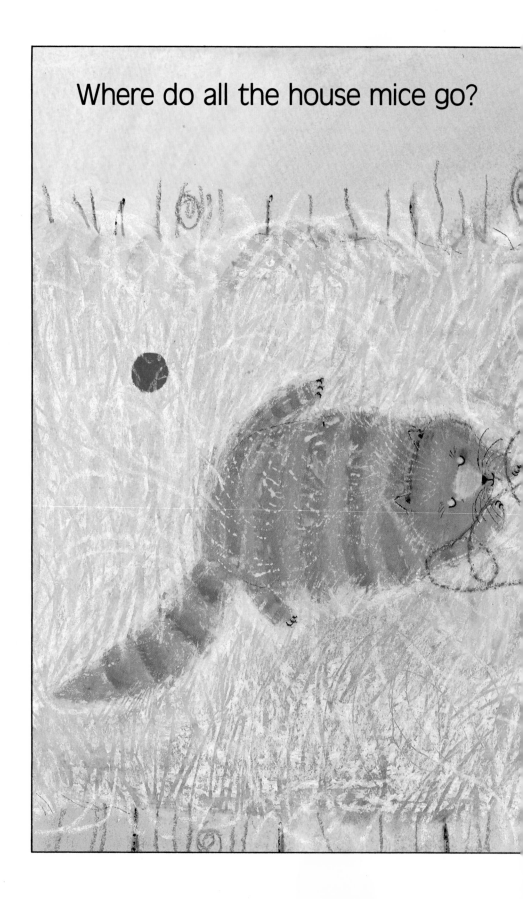

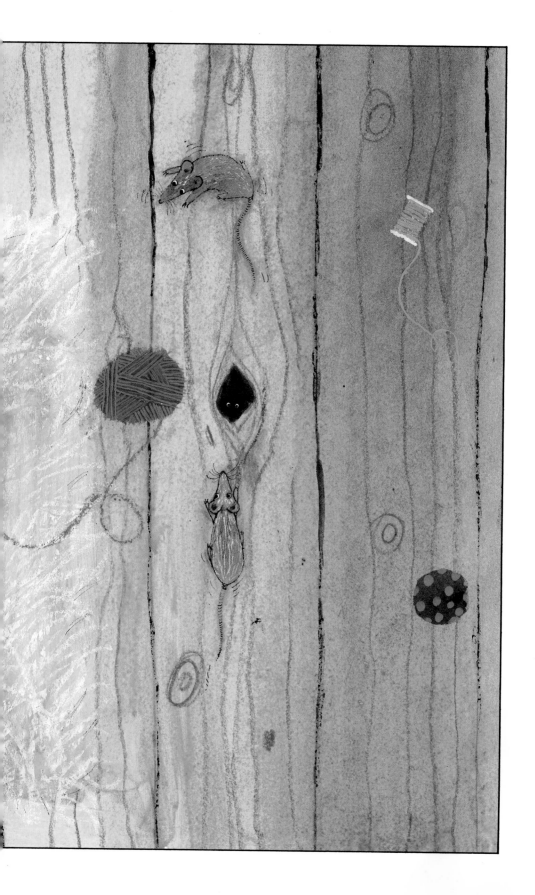

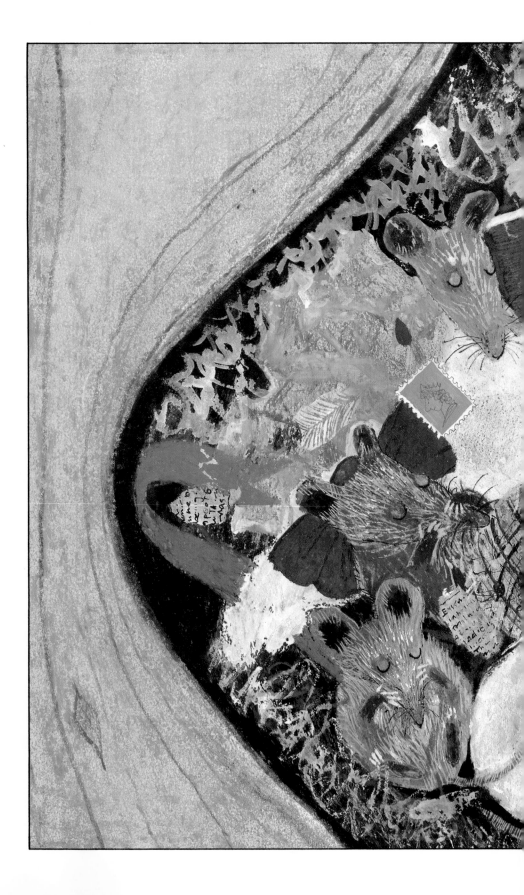

They curl up very
tight in nests made
of odds and ends.

Where do all the fish go?

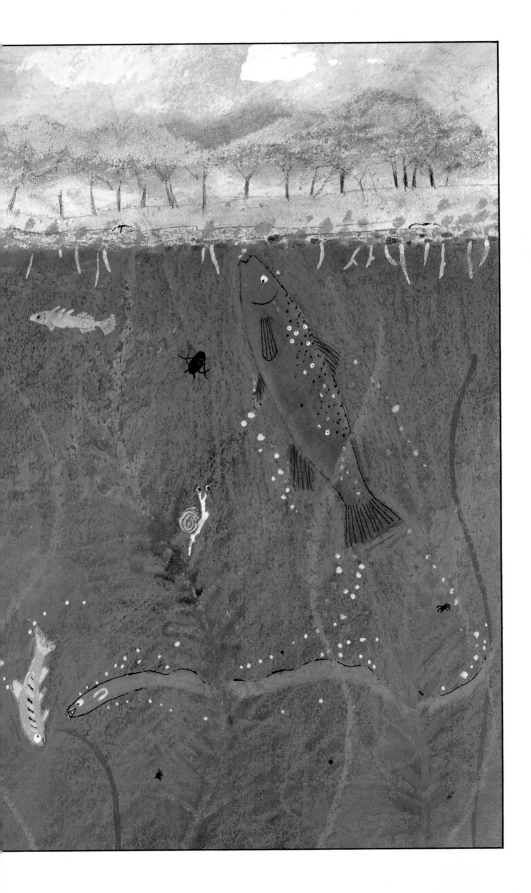

They swim to the bottom
of the pond where the water
is warmer.

And where do all the people go?

They go home!